6 Full-Length Georgia Milestones Assessment System Grade 6 Math Practice Tests

Extra Test Prep to Help Ace the GMAS Grade 6 Math Test

By

Michael Smith & Reza Nazari

6 Full-Length Georgia Milestones Assessment System Grade 6 Math Practice Tests

Published in the United State of America By

The Math Notion

Web: WWW.MathNotion.Com

Email: info@Mathnotion.com

Copyright © 2020 by the Math Notion. All rights reserved. No part of this publication may be reproduced, stored in a retrieval system, or transmitted in any form or by any means, electronic, mechanical, photocopying, recording, scanning, or otherwise, except as permitted under Section 107 or 108 of the 1976 United States Copyright Ac, without permission of the author.

All inquiries should be addressed to the Math Notion.

About the Author

Michael Smith has been a math instructor for over a decade now. He holds a master's degree in Management. Since 2006, Michael has devoted his time to both teaching and developing exceptional math learning materials. As a Math instructor and test prep expert, Michael has worked with thousands of students. He has used the feedback of his students to develop a unique study program that can be used by students to drastically improve their math score fast and effectively.

- **– SAT Math Practice Book**
- **– ACT Math Practice Book**
- **– GRE Math Practice Book**
- **– Common Core Math Practice Book**
- **–many Math Education Workbooks, Exercise Books and Study Guides**

As an experienced Math teacher, Mr. Smith employs a variety of formats to help students achieve their goals: He tutors online and in person, he teaches students in large groups, and he provides training materials and textbooks through his website and through Amazon.

You can contact Michael via email at:

info@Mathnotion.com

Prepare for the Georgia Milestones Assessment System Grade 6 Math test with a perfect practice book!

The surest way to practice your GMAS Math test-taking skills is with simulated exams. This comprehensive practice book with 6 full length and realistic GMAS Math practice tests help you measure your exam readiness, find your weak areas, and succeed on the GMAS Math test. The detailed answers and explanations for each GMAS Math question help you master every aspect of the GMAS Math.

6 Full-length Georgia Milestones Assessment System Grade 6 Math Practice Tests is a prestigious resource to help you succeed on the GMAS Math test. This perfect practice book features:

- Content 100% aligned with the GMAS test
- Six full-length GMAS Math practice tests similar to the actual test in length, format, question types, and degree of difficulty
- Detailed answers and explanations for the GMAS Math practice questions
- Written by GMAS Math top instructors and experts

After completing this hands-on exercise book, you will gain confidence, strong foundation, and adequate practice to succeed on the GMAS Math test.

WWW.MathNotion.COM

... So Much More Online!

✓ FREE Math Lessons

✓ More Math Learning Books!

✓ Mathematics Worksheets

✓ Online Math Tutors

For a PDF Version of This Book

Please Visit WWW.MathNotion.com

Contents

GMAS Math Practice Tests 9

GMAS Grade 6 Mathematics Reference Sheet 11

Georgia Milestones Assessment System Practice Test 1 13
   Session 1 14
   Session 2 18

Georgia Milestones Assessment System Practice Test 2 23
   Session 1 24
   Session 2 28

Georgia Milestones Assessment System Practice Test 3 33
   Session 1 34
   Session 2 38

Georgia Milestones Assessment System Practice Test 4 43
   Session 1 44
   Session 2 48

Georgia Milestones Assessment System Practice Test 5 53
   Session 1 54
   Session 2 58

Georgia Milestones Assessment System Practice Test 6 63
   Session 1 64
   Session 2 68

Answer Keys 73

Answers and Explanations 77
   Practice Test 1 79
   Practice Test 2 83
   Practice Test 3 87
   Practice Test 4 91

Practice Test 5 95
Practice Test 6 99

GMAS Math Practice Tests

Time to Test

Time to refine your skill with a practice examination

Take a REAL GMAS Mathematics test to simulate the test day experience. After you've finished, score your test using the answer key.

Before You Start

- You'll need a pencil and scratch papers to take the test.
- For this practice test, don't time yourself. Spend time as much as you need.
- It's okay to guess. You won't lose any points if you're wrong.
- After you've finished the test, review the answer key to see where you went wrong.

Calculators are not permitted for GMAS Tests

Good Luck!

GMAS Grade 6 Mathematics Reference Sheet

Conversions:

LENGTH

Customary	Metric
1 mile (mi) = 1,760 yards (yd)	1 kilometer (km) = 1,000 meters (m)
1 yard (yd) = 3 feet (ft)	1 meter (m) = 100 centimeters (cm)
1 foot (ft) = 12 inches (in.)	1 centimeter (cm) = 10 millimeters (mm)

VOLUME AND CAPACITY

Customary	Metric
1 gallon (gal) = 4 quarts (qt)	1 liter (L) = 1,000 milliliters (mL)
1 quart (qt) = 2 pints (pt.)	
1 pint (pt.) = 2 cups (c)	
1 cup (c) = 8 fluid ounces (Fl oz)	

WEIGHT AND MASS

Customary	Metric
1 ton (T) = 2,000 pounds (lb.)	1 kilogram (kg) = 1,000 grams (g)
1 pound (lb.) = 16 ounces (oz)	1 gram (g) = 1,000 milligrams (mg)

Formulas:

Area

Triangle	$A = \frac{1}{2}bh$
Rectangle or Parallelogram	$A = bh$
Trapezoid	$A = \frac{1}{2}h(b_1 + b_2)$

Volume

Rectangular Prism	$V = Bh$

Georgia Milestones Assessment

System Practice Test 1

Mathematics

GRADE 6

Administered *Month Year*

Session 1

- ❖ Calculators are permitted for this practice test.
- ❖ Time for Session 1: 85 Minutes

1) If $x = -3$, which of the following equations is true?

   A. $x(4x - 1) = 35$

   B. $2(12 - x^2) = -6$

   C. $4(-2x + 4) = 42$

   D. $x(-7x - 12) = -27$

2) What is the perimeter of the following shape? (it's a right triangle)

   A. 5 cm

   B. 16 cm

   C. 24 cm

   D. 12 cm

   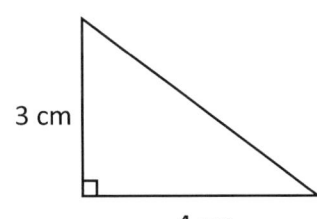

3) 60 is what percent of 25?

   Write your answer in the box below.

   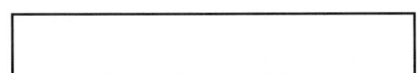

4) Which of the following expressions has a value of -8?

   A. $-5 + (-18 \div 3) + \frac{-6}{5} \times 5$

   B. $2 \times (-10) + (-3) \times 4$

   C. $(-2) + 14 \times 3 \div (-7)$

   D. $(-4) \times (-8) + 5$

5) A football team won exactly 70% of the games it played during last session. Which of the following could be the total number of games the team played last season?

   A. 61

   B. 50

   C. 42

   D. 25

6) 6 less than twice a positive integer is 80. What is the integer?

   A. 80

   B. 43

   C. 74

   D. 40

7) Which of the following expressions has the greatest value?

   A. $5^3 - 4^3$

   B. $2^5 - 2^2$

   C. $3^4 - 5^2$

   D. $4^4 - 15^2$

8) The diameter of a circle is 3π. What is the area of the circle?

   A. $9\pi^2$

   B. $\frac{3\pi^2}{2}$

   C. $\frac{9\pi^3}{4}$

   D. $\frac{\pi^3}{4}$

9) Elise has x apples. Alvin has 45 apples, which is 20 apples less than number of apples Elise owns. If Baron has $\frac{1}{5}$ times as many apples as Elise has. How many apples does Baron have?

   A. 11

   B. 45

   C. 20

   D. 13

10) Find the perimeter of shape in the following figure? (all angles are right angles)

Write your answer in the box below.

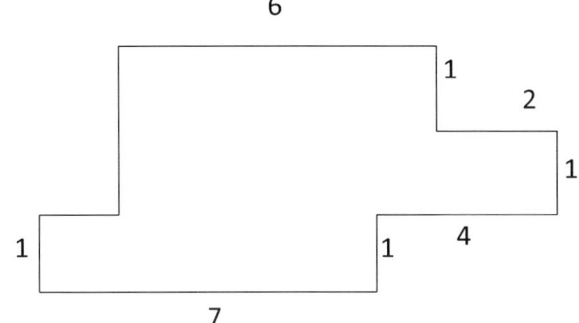

Session 2

- ❖ Calculators are permitted for this practice test.
- ❖ Time for Session 2: 85 Minutes

11) Car A travels 341.26 km at a given time, while car B travels 2.2 times the distance car A travels at the same time. What is the distance car B travels during that time?

   A. 650.7 km

   B. 453.5 km

   C. 341.2 km

   D. 750.8 km

12) What is the probability of choosing a month starts with A in a year?

   A. 1

   B. $\frac{2}{3}$

   C. $\frac{1}{2}$

   D. $\frac{1}{6}$

13) What are the values of mode and median in the following set of numbers?

   1, 2, 2, 4, 4, 5, 6, 3, 1, 1, 3

   A. Mode: 1, 2 Median: 3

   B. Mode: 1, 3 Median: 2

   C. Mode: 2, Median: 2

   D. Mode: 1, Median: 3

14) If point A placed at $-\frac{16}{4}$ on a number line, which of the following points has a distance equal to 6 from point A?

   A. -10

   B. 2

   C. -3

   D. A and B

15) The ratio of pens to pencils in a box is 5 to 7. If there are 144 pens and pencils in the box altogether, how many more pens should be put in the box to make the ratio of pens to pencils 1: 1?

   Write your answer in the box below.

   [ ]

16) Which of the following shows the numbers in increasing order?

   A. $\frac{4}{2}, \frac{7}{4}, \frac{9}{12}, \frac{35}{8}$

   B. $\frac{9}{12}, \frac{7}{4}, \frac{4}{2}, \frac{35}{8}$

   C. $\frac{9}{12}, \frac{4}{2}, \frac{7}{4}, \frac{35}{8}$

   D. $\frac{35}{8}, \frac{7}{4}, \frac{9}{12}, \frac{4}{2}$

17) If $3x - 2 = 16$, what is the value of $3x + 2$?

   A. 28

   B. 29

   C. 9

   D. 15

18) $5(1.153) - 2.126 = \cdots?$

Write your answer in the box below.

19) In the following triangle find α.

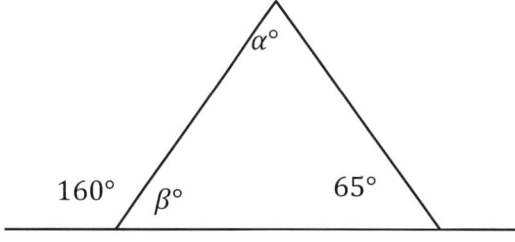

Write your answer in the box below.

20) A shaft rotates 360 times in 9 seconds. How many times does it rotate in 16 seconds?

   A. 540

   B. 640

   C. 360

   D. 200

Practice Test 1

This is the End of this Section

Georgia Milestones Assessment

System Practice Test 2

Mathematics

GRADE 6

Administered *Month Year*

Session 1

- ❖ Calculators are permitted for this practice test.
- ❖ Time for Session 1: 85 Minutes

1) Martin earns $30 an hour. Which of the following inequalities represents the amount of time Martin needs to work per day to earn at least $250 per day?

   A. $30t \geq 250$

   B. $30t \leq 250$

   C. $30 + t \geq 250$

   D. $30 + t \leq 250$

2) What is the value of the expression $4(3x - 2y) + (4 - 5x)^2$, when $x = 1$ and $y = -2$?

   Write your answer in the box below.

   [ ]

3) Round $\frac{415}{9}$ to the nearest tenth.

   A. 46

   B. 46.3

   C. 46.1

   D. 46.6

4) Which expression is equivalent to $6(10x - 12)$?

   A. -60

   B. $-60x$

   C. $60x - 72$

   D. $60x - 60$

5) A chemical solution contains 5% alcohol. If there is 40 ml of alcohol, what is the volume of the solution?

   A. 700 ml

   B. 450 ml

   C. 800 ml

   D. 2,000 ml

6) Which ordered pair describes point A that is shown below?

   A. $(3, -2)$

   B. $(2, -3)$

   C. $(-3, 2)$

   D. $(-2, -3)$

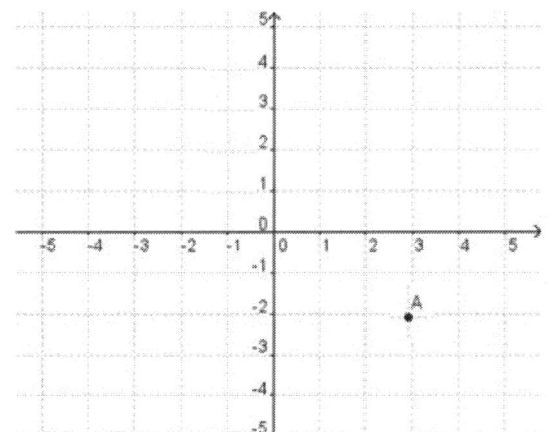

7) To produce a special concrete, for every 14 kg of cement, 2 liters of water is required. Which of the following ratios is the same as the ratio of cement to liters of water?

   A. 84: 12

   B. 20: 8

   C. 28: 14

   D. 4: 16

8) Find the opposite of the numbers 16, 0.

Write your answer in the box below.

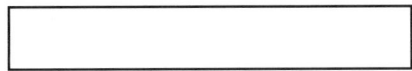

9) What is the value of x in the following equation?

$$-50 = 85 - x$$

A. 135

B. −135

C. 75

D. −75

10) Which of the following graphs represents the following inequality?

$$-2 \leq 2x - 4 < 8$$

A.

B.

C.

D.

Session 2

❖ Calculators are permitted for this practice test.

❖ Time for Session 2: 85 Minutes

11) The ratio of boys to girls in a school is 5:7. If there are 900 students in the school, how many boys are in the school?

   A. 750

   B. 435

   C. 654

   D. 375

12) $(75 + 5) \div 16$ is equivalent to …

   A. $80 \div 3.4$

   B. $\frac{55}{16} + 5$

   C. $(2 \times 2 \times 4 \times 5) \div (4 \times 4)$

   D. $(2 \times 2 \times 2 \times 5) \div 4 + 4$

13) What is the equation of a line that passes through points (0, 5) and (4, 9)?

   A. $y = x$

   B. $y = x + 5$

   C. $y = 2x + 5$

   D. $y = 2x - 5$

14) What is the volume of a box with the following dimensions? Height = 5cm

Width = 6 cm   Length = 8 cm

Write your answer in the box below.

☐

15) Anita's trick–or–treat bag contains 16 pieces of chocolate, 18 suckers, 14 pieces of gum, 12 pieces of licorice. If she randomly pulls a piece of candy from her bag, what is the probability of her pulling out a piece of sucker?

A. $\frac{1}{18}$

B. $\frac{3}{10}$

C. $\frac{6}{10}$

D. $\frac{18}{18} = 1$

16) What is the lowest common multiple of 18 and 30?

A. 90

B. 40

C. 30

D. 45

17) Which statement is true about all rectangles?

A: Both diagonals have equal measure.

B: All sides are congruent.

C: Both diagonals are perpendicular.

D: All the statements are true

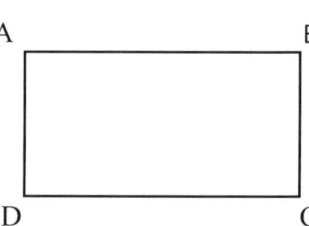

18) Which of the following lists shows the fractions in order from least to greatest?

$$\frac{3}{4}, \frac{4}{5}, \frac{11}{5}, \frac{26}{29}$$

A. $\frac{26}{29}, \frac{4}{5}, \frac{3}{4}, \frac{11}{5}$

B. $\frac{4}{5}, \frac{26}{29}, \frac{11}{5}, \frac{3}{4}$

C. $\frac{3}{4}, \frac{4}{5}, \frac{26}{29}, \frac{11}{5}$

D. $\frac{26}{29}, \frac{4}{5}, \frac{11}{5}, \frac{3}{4}$

19) Which statement about 3 multiplied by $\frac{5}{7}$ must be true?

A. The product is between 1 and 2

B. The product is greater than 3

C. The product is equal to $\frac{77}{25}$

D. The product is between 2 and 2.3

20) If the area of the following rectangular ABCD is 160, and E is the midpoint of AB, what is the area of the shaded part?

Write your answer in the box below.

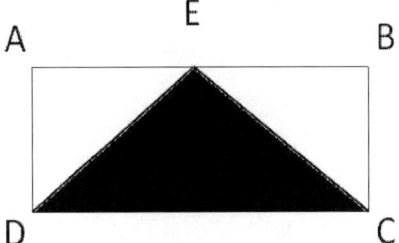

Practice Test 2

This is the End of this Section

Georgia Milestones Assessment System Practice Test 3

Mathematics

GRADE 6

Administered *Month Year*

Session 1

- ❖ Calculators are permitted for this practice test.
- ❖ Time for Session 1: 85 Minutes

GMAS Math Practice Tests – Grade 6

1) If $x = -3$, which of the following equations is true?

   A. $x(2x - 4) = 18$

   B. $5(12 - x^2) = 15$

   C. $2(-x + 5) = 21$

   D. $x(-2x - 11) = -19$

2) What is the perimeter of the following shape? (it's a right triangle)

   A. 18 cm

   B. 16 cm

   C. 14 cm

   D. 12 cm

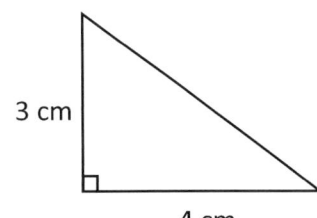

3) 30 is what percent of 12?

   Write your answer in the box below.

   [ ]

4) Which of the following expressions has a value of -12?

   A. $-4 + (-18 \div 3) + \frac{-9}{5} \times 5$

   B. $6 \times (-4) + (-3) \times 4$

   C. $(-6) + 10 \times 3 \div (-5)$

   D. $(-4) \times (-6) + 11$

5) A football team won exactly 20% of the games it played during last session. Which of the following could be the total number of games the team played last season?

   A. 46

   B. 35

   C. 52

   D. 28

6) 5 less than triple a positive integer is 70. What is the integer?

   A. 55

   B. 25

   C. 75

   D. 65

7) Which of the following expressions has the greatest value?

   A. $7^3 - 6^3$

   B. $2^5 - 5^2$

   C. $3^4 - 7^2$

   D. $13^2 - 5^3$

8) The diameter of a circle is 3π. What is the area of the circle?

A. $\dfrac{9\pi^3}{4}$

B. $\dfrac{9\pi^2}{2}$

C. $\dfrac{6\pi^3}{4}$

D. $9\pi^3$

9) Elise has x apples. Alvin has 40 apples, which is 20 apples less than number of apples Elise owns. If Baron has $\dfrac{1}{3}$ times as many apples as Elise has. How many apples does Baron have?

A. 45

B. 35

C. 20

D. 30

10) Find the perimeter of shape in the following figure? (all angles are right angles)

Write your answer in the box below.

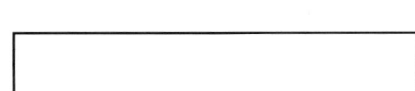

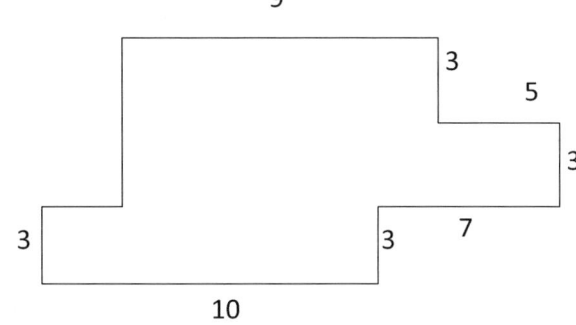

Session 2

❖ Calculators are permitted for this practice test.

❖ Time for Session 2: 85 Minutes

11) Car A travels 250.50 km at a given time, while car B travels 2.5 times the distance car A travels at the same time. What is the distance car B travels during that time?

A. 266.50 km

B. 256.25 km

C. 566.25 km

D. 626.25 km

12) What is the probability of choosing a month starts with O in a year?

A. 1

B. $\frac{1}{2}$

C. $\frac{1}{6}$

D. $\frac{1}{12}$

13) What are the values of mode and median in the following set of numbers?

$$3, 5, 4, 4, 7, 6, 8, 5, 6, 9, 5$$

A. Mode: 5, Median: 9

B. Mode: 8, 3 Median: 8

C. Mode: 5, Median: 6

D. Mode: 5, Median: 5

14) If point A placed at $-\frac{21}{7}$ on a number line, which of the following points has a distance equal to 6 from point A?

   A. -9

   B. -3

   C. 3

   D. A and C

15) The ratio of pens to pencils in a box is 2 to 5. If there are 91 pens and pencils in the box altogether, how many more pens should be put in the box to make the ratio of pens to pencils 1: 1?

   Write your answer in the box below.

   ☐

16) Which of the following shows the numbers in increasing order?

   A. $\frac{13}{4}, \frac{8}{3}, \frac{7}{9}, \frac{38}{6}$

   B. $\frac{8}{3}, \frac{7}{9}, \frac{13}{4}, \frac{38}{6}$

   C. $\frac{7}{9}, \frac{8}{3}, \frac{13}{4}, \frac{38}{6}$

   D. $\frac{38}{6}, \frac{8}{3}, \frac{7}{9}, \frac{13}{4}$

17) If $6x - 2 = 22$, what is the value of $5x + 5$?

   A. 15

   B. 25

   C. 10

   D. 40

18) $4(1.75) - 3.75 = \cdots?$

Write your answer in the box below.

☐

19) In the following triangle find α.

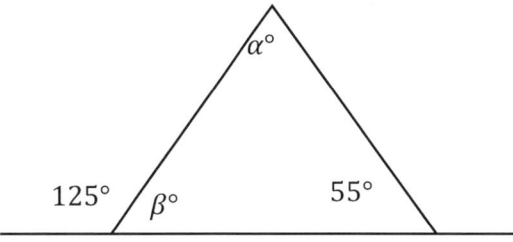

Write your answer in the box below.

☐

20) A shaft rotates 330 times in 9 seconds. How many times does it rotate in 12 seconds?

A. 420

B. 440

C. 220

D. 140

Practice Test 3

This is the End of this Section

//GMAS Math Practice Tests – Grade 6

Georgia Milestones Assessment System Practice Test 4

Mathematics

GRADE 6

Administered *Month Year*

Session 1

- ❖ Calculators are permitted for this practice test.
- ❖ Time for Session 1: 85 Minutes

1) Martin earns $50 an hour. Which of the following inequalities represents the amount of time Martin needs to work per day to earn at least $520 per day?

   A. $50t \geq 520$

   B. $50t \leq 520$

   C. $50 + t \geq 520$

   D. $50 + t \leq 520$

2) What is the value of the expression $2(3x - y) + (7 - 3x)^2$, when $x = 1$ and $y = -2$?

   Write your answer in the box below.

3) Round $\frac{155}{6}$ to the nearest tenth.

   A. 26.5

   B. 27

   C. 26

   D. 25.5

4) Which expression is equivalent to $9(10x - 6)$?

   A. -90

   B. $-90x$

   C. $90x - 54$

   D. $90x - 60$

5) A chemical solution contains 7% alcohol. If there is 42 ml of alcohol, what is the volume of the solution?

A. 460 ml

B. 660 ml

C. 600 ml

D. 1,600 ml

6) Which ordered pair describes point A that is shown below?

A. $(3, -3)$

B. $(-3, 3)$

C. $(-3, -3)$

D. $(3, 3)$

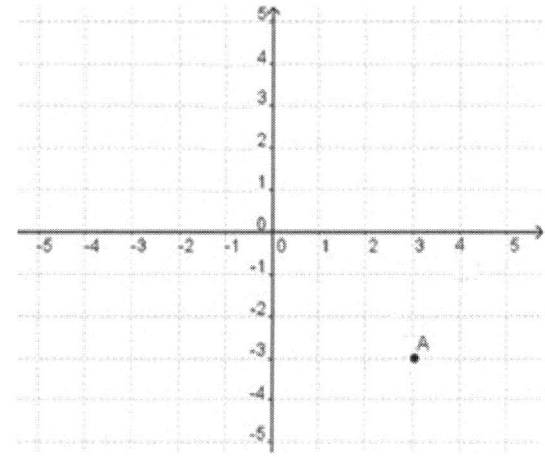

7) To produce a special concrete, for every 30 kg of cement, 5 liters of water is required. Which of the following ratios is the same as the ratio of cement to liters of water?

A. 90: 15

B. 80: 20

C. 45: 15

D. 90: 25

8) Find the opposite of the numbers 17, 3.

Write your answer in the box below.

9) What is the value of x in the following equation?

$$-32 = 74 - x$$

A. 106

B. -106

C. 85

D. -85

10) Which of the following graphs represents the following inequality?

$$1 \leq 2x - 5 < 11$$

A.
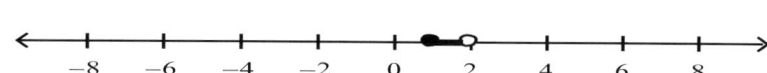

B.

C.
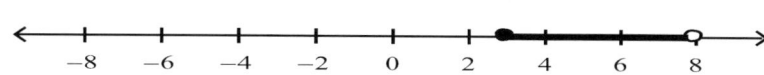

D.

Session 2

- ❖ Calculators are permitted for this practice test.
- ❖ Time for Session 2: 85 Minutes

11) The ratio of boys to girls in a school is 4:5. If there are 450 students in the school, how many boys are in the school?

   A. 250

   B. 420

   C. 210

   D. 200

12) $(80 + 4) \div 35$ is equivalent to …

   A. $80 \div 3.5$

   B. $\frac{80}{35} + 4$

   C. $(2 \times 2 \times 3 \times 7) \div (5 \times 7)$

   D. $(2 \times 2 \times 3 \times 5) \div 5 + 7$

13) What is the equation of a line that passes through points (3, 2) and (4, 7)?

   A. $y = -5x$

   B. $y = 5x - 13$

   C. $y = 5x + 13$

   D. $y = x + 5$

14) What is the volume of a box with the following dimensions? Height = 6cm

Width = 6 cm   Length = 8 cm

Write your answer in the box below.

☐

15) Anita's trick–or–treat bag contains 16 pieces of chocolate, 25 suckers, 20 pieces of gum, 19 pieces of licorice. If she randomly pulls a piece of candy from her bag, what is the probability of her pulling out a piece of sucker?

A. $\frac{1}{25}$

B. $\frac{5}{16}$

C. $\frac{7}{16}$

D. $\frac{25}{25} = 1$

16) What is the lowest common multiple of 15 and 50?

A. 150

B. 50

C. 200

D. 75

17) Which statement is true about all square?

A. Both diagonals have equal measure.

B. All sides are congruent.

C. Both diagonals are perpendicular.

D. All the statements are true

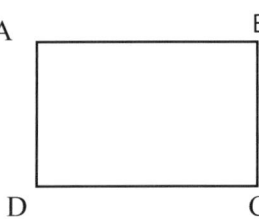

18) Which of the following lists shows the fractions in order from least to greatest?

$$\frac{6}{7}, \frac{2}{5}, \frac{23}{6}, \frac{19}{17}$$

A. $\frac{19}{17}, \frac{6}{7}, \frac{2}{5}, \frac{23}{6}$

B. $\frac{6}{7}, \frac{19}{17}, \frac{23}{6}, \frac{2}{5}$

C. $\frac{2}{5}, \frac{6}{7}, \frac{19}{17}, \frac{23}{6}$

D. $\frac{19}{17}, \frac{6}{7}, \frac{23}{6}, \frac{2}{5}$

19) Which statement about 3 multiplied by $\frac{6}{7}$ must be true?

A. The product is between 4 and 5

B. The product is greater than 3

C. The product is equal to $\frac{35}{15}$

D. The product is between 2 and 3

20) If the area of the following rectangular ABCD is 240, and E is the midpoint of AB, what is the area of the shaded part?

Write your answer in the box below.

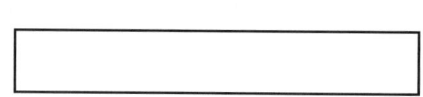

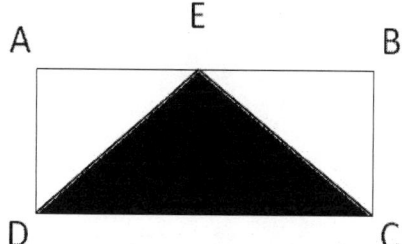

Practice Test 4
This is the End of this Section

Georgia Milestones Assessment System Practice Test 5

Mathematics

GRADE 6

Administered *Month Year*

Session 1

- ❖ Calculators are permitted for this practice test.
- ❖ Time for Session 1: 85 Minutes

1) If $x = -2$, which of the following equations is true?

   A. $x(3x - 2) = 32$

   B. $2(10 - x^2) = 12$

   C. $3(-x + 3) = 18$

   D. $x(-4x - 10) = -17$

2) What is the perimeter of the following shape? (it's a right triangle)

   A. 15 cm

   B. 26 cm

   C. 34 cm

   D. 24 cm

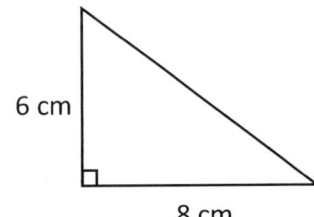

3) 45 is what percent of 15?

   Write your answer in the box below.

4) Which of the following expressions has a value of -9?

   A. $-3 + (-16 \div 4) + \frac{-8}{7} \times 7$

   B. $3 \times (-9) + (-2) \times 5$

   C. $(-3) + 12 \times 4 \div (-8)$

   D. $(-5) \times (-9) + 36$

5) A football team won exactly 60% of the games it played during last session. Which of the following could be the total number of games the team played last season?

   A. 56

   B. 25

   C. 32

   D. 48

6) 8 less than twice a positive integer is 90. What is the integer?

   A. 50

   B. 49

   C. 34

   D. 45

7) Which of the following expressions has the greatest value?

   A. $6^3 - 5^3$

   B. $3^4 - 4^2$

   C. $4^3 - 6^2$

   D. $6^3 - 12^2$

8) The diameter of a circle is 8π. What is the area of the circle?

   A. $16\pi^3$

   B. $\frac{13\pi^2}{2}$

   C. $\frac{15\pi^3}{4}$

   D. $\frac{3\pi^3}{4}$

9) Elise has x apples. Alvin has 50 apples, which is 30 apples less than number of apples Elise owns. If Baron has $\frac{1}{4}$ times as many apples as Elise has. How many apples does Baron have?

   A. 15

   B. 25

   C. 20

   D. 10

10) Find the perimeter of shape in the following figure? (all angles are right angles)

Write your answer in the box below.

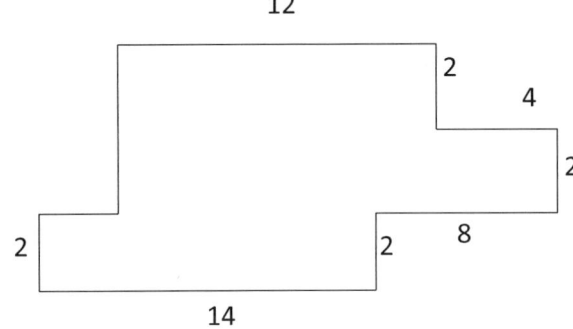

Session 2

- ❖ Calculators are permitted for this practice test.
- ❖ Time for Session 2: 85 Minutes

11) Car A travels 241.15 km at a given time, while car B travels 3.2 times the distance car A travels at the same time. What is the distance car B travels during that time?

   A. 659.77 km

   B. 753.65 km

   C. 381.82 km

   D. 771.68 km

12) What is the probability of choosing a month starts with M in a year?

   A. 1

   B. $\frac{1}{3}$

   C. $\frac{1}{4}$

   D. $\frac{1}{6}$

13) What are the values of mode and median in the following set of numbers?

   $$1, 5, 2, 6, 6, 5, 6, 3, 4, 8, 3$$

   A. Mode: 1, Median: 8

   B. Mode: 5, 3 Median: 4

   C. Mode: 5, Median: 3

   D. Mode: 6, Median: 5

14) If point A placed at $-\frac{12}{3}$ on a number line, which of the following points has a distance equal to 8 from point A?

A. -12

B. 4

C. -6

D. A and B

15) The ratio of pens to pencils in a box is 3 to 4. If there are 98 pens and pencils in the box altogether, how many more pens should be put in the box to make the ratio of pens to pencils 1: 1?

Write your answer in the box below.

16) Which of the following shows the numbers in increasing order?

A. $\frac{5}{2}, \frac{9}{3}, \frac{8}{15}, \frac{45}{8}$

B. $\frac{8}{15}, \frac{9}{3}, \frac{5}{2}, \frac{45}{8}$

C. $\frac{8}{15}, \frac{5}{2}, \frac{9}{3}, \frac{45}{8}$

D. $\frac{45}{8}, \frac{9}{3}, \frac{8}{15}, \frac{5}{2}$

17) If $4x - 3 = 17$, what is the value of $4x + 3$?

A. 18

B. 23

C. 19

D. 12

18) $3(2.313) - 1.546 = \cdots ?$

Write your answer in the box below.

19) In the following triangle find α.

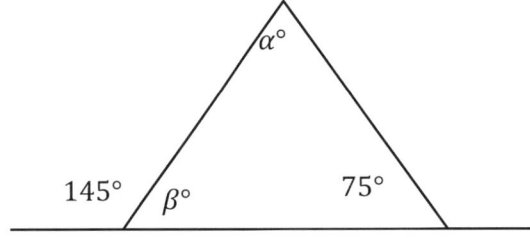

Write your answer in the box below.

20) A shaft rotates 440 times in 11 seconds. How many times does it rotate in 15 seconds?

A. 500

B. 600

C. 300

D. 700

Practice Test 5

This is the End of this Section

Georgia Milestones Assessment System Practice Test 6

Mathematics

GRADE 6

Administered *Month Year*

Session 1

- ❖ Calculators are permitted for this practice test.
- ❖ Time for Session 1: 85 Minutes

1) Martin earns $40 an hour. Which of the following inequalities represents the amount of time Martin needs to work per day to earn at least $450 per day?

   A. $40t \geq 450$

   B. $40t \leq 450$

   C. $40 + t \geq 450$

   D. $40 + t \leq 450$

2) What is the value of the expression $2(2x - 2y) + (3 - 4x)^2$, when $x = 2$ and $y = -1$?

   Write your answer in the box below.

   ☐

3) Round $\frac{369}{7}$ to the nearest tenth.

   A. 52

   B. 56.3

   C. 52.7

   D. 52.6

4) Which expression is equivalent to $8(11x - 9)$?

   A. -88

   B. $-88x$

   C. $88x - 72$

   D. $88x - 70$

5) A chemical solution contains 6% alcohol. If there is 36 ml of alcohol, what is the volume of the solution?

A. 750 ml

B. 650 ml

C. 600 ml

D. 3,000 ml

6) Which ordered pair describes point A that is shown below?

A. $(1, -3)$

B. $(-1, 3)$

C. $(-1, 3)$

D. $(-1, -3)$

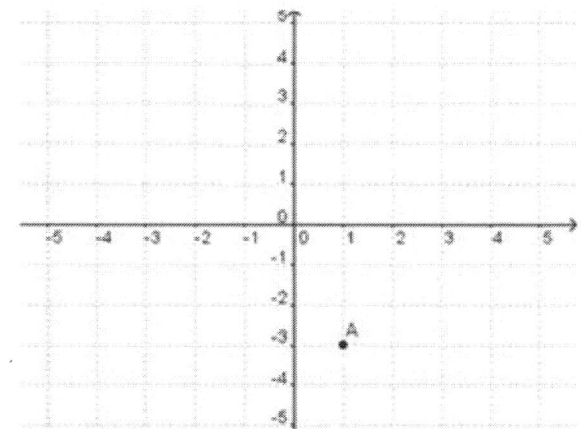

7) To produce a special concrete, for every 12 kg of cement, 4 liters of water is required. Which of the following ratios is the same as the ratio of cement to liters of water?

A. 36: 12

B. 36: 6

C. 24: 12

D. 12: 6

8) Find the opposite of the numbers 22, 0.

Write your answer in the box below.

9) What is the value of x in the following equation?

$$-42 = 94 - x$$

A. 136

B. -136

C. 95

D. -95

10) Which of the following graphs represents the following inequality?

$$-3 \leq 3x - 6 < 9$$

A.
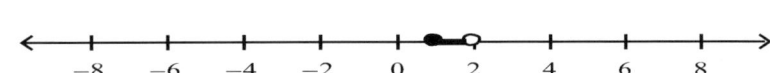

B.

C.
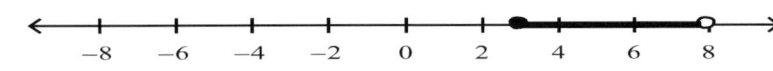

D.

Session 2

❖ Calculators are permitted for this practice test.

❖ Time for Session 2: 85 Minutes

GMAS Math Practice Tests – Grade 6

11) The ratio of boys to girls in a school is 3:4. If there are 700 students in the school, how many boys are in the school?

    A. 450

    B. 400

    C. 610

    D. 300

12) $(45 + 5) \div 25$ is equivalent to …

    A. $50 \div 3.4$

    B. $\frac{50}{15} + 5$

    C. $(2 \times 5 \times 5) \div (5 \times 5)$

    D. $(2 \times 5 \times 5) \div 5 + 5$

13) What is the equation of a line that passes through points $(1, 2)$ and $(3, 4)$?

    A. $y = -x$

    B. $y = x + 1$

    C. $y = 2x + 1$

    D. $y = 2x - 1$

14) What is the volume of a box with the following dimensions? Height = 3cm

Width = 5 cm   Length = 10 cm

Write your answer in the box below.

15) Anita's trick–or–treat bag contains 13 pieces of chocolate, 20 suckers, 15 pieces of gum, 17 pieces of licorice. If she randomly pulls a piece of candy from her bag, what is the probability of her pulling out a piece of sucker?

A. $\frac{1}{20}$

B. $\frac{4}{13}$

C. $\frac{6}{13}$

D. $\frac{20}{20} = 1$

16) What is the lowest common multiple of 24 and 40?

A. 120

B. 80

C. 60

D. 85

17) Which statement is true about all square?

A. Both diagonals have equal measure.

B. All sides are congruent.

C. Both diagonals are perpendicular.

D. All the statements are true

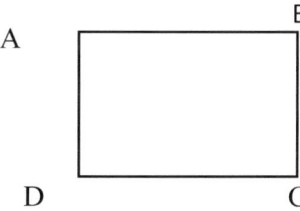

18) Which of the following lists shows the fractions in order from least to greatest?

$$\frac{3}{5}, \frac{5}{7}, \frac{13}{6}, \frac{27}{30}$$

A. $\frac{27}{30}, \frac{5}{7}, \frac{3}{5}, \frac{13}{6}$

B. $\frac{5}{7}, \frac{27}{30}, \frac{13}{6}, \frac{3}{5}$

C. $\frac{3}{5}, \frac{5}{7}, \frac{27}{30}, \frac{13}{6}$

D. $\frac{27}{30}, \frac{5}{7}, \frac{13}{6}, \frac{3}{5}$

19) Which statement about 4 multiplied by $\frac{7}{11}$ must be true?

A. The product is between 1.5 and 2

B. The product is greater than 4

C. The product is equal to $\frac{77}{25}$

D. The product is between 2 and 3

20) If the area of the following rectangular ABCD is 120, and E is the midpoint of AB, what is the area of the shaded part?

Write your answer in the box below.

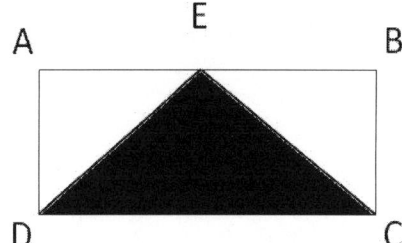

Practice Test 6
This is the End of this Section

Answer Keys
Georgia Milestones Assessment System Practice Tests

❋ Now, it's time to review your results to see where you went wrong and what areas you need to improve!

Practice Test - 1				Practice Test - 2			
1	D	11	D	1	A	11	D
2	D	12	D	2	29	12	C
3	240%	13	D	3	C	13	B
4	C	14	D	4	C	14	240
5	B	15	24	5	C	15	B
6	B	16	B	6	A	16	A
7	A	17	B	7	A	17	A
8	C	18	3.639	8	−16,0	18	C
9	D	19	95°	9	A	19	D
10	28	20	B	10	D	20	80

Practice Test - 3

1	B	11	D
2	D	12	D
3	250%	13	D
4	C	14	D
5	B	15	39
6	B	16	C
7	A	17	B
8	A	18	3.25
9	C	19	70°
10	52	20	B

Practice Test - 4

1	A	11	D
2	26	12	C
3	C	13	B
4	C	14	288
5	C	15	B
6	A	16	A
7	A	17	D
8	−17, −3	18	C
9	A	19	D
10	C	20	120

Practice Test - 5

1	B	11	D
2	D	12	D
3	300%	13	D
4	C	14	D
5	B	15	14
6	B	16	C
7	A	17	B
8	A	18	5.393
9	C	19	70°
10	56	20	B

Practice Test - 6

1	A	11	D
2	37	12	C
3	C	13	B
4	C	14	150
5	C	15	B
6	A	16	A
7	A	17	D
8	−22,0	18	C
9	A	19	D
10	D	20	60

Answers and Explanations

Practice Test 1

Georgia Milestones Assessment

System - Mathematics

Answers and Explanations

1) Answer: D

Plugin the value of x in the equations. $x = -3$, then:

A. $x(4x - 1) = 35 \rightarrow -3(4(-3) - 1) = -3(-12 - 1) = -3(-13) = 39 \neq 35$

B. $2(12 - x^2) = -6 \rightarrow 2(12 - (-3)^2) = 2(12 - 9) = 2(3) \neq -6$

C. $4(-2x + 4) = 42 \rightarrow 4(-2(-3) + 4) = 4(6 + 4) = 40 \neq 42$

D. $x(-7x - 12) = -27 \rightarrow -3(-7(-3) - 12 = -3(21 - 12) = -27 = -27$

2) Answer: D

Use Pythagorean theorem to find the hypotenuse of the triangle.

$a^2 + b^2 = c^2 \rightarrow 3^2 + 4^2 = c^2 \rightarrow 9 + 16 = c^2 \rightarrow 25 = c^2 \rightarrow c = 5$

The perimeter of the triangle is: $3 + 4 + 5 = 12$

3) Answer: 240%

Use percent formula: Part $= \frac{\text{percent}}{100} \times$ whole

$60 = \frac{\text{percent}}{100} \times 25 \Rightarrow 60 = \frac{\text{percent} \times 25}{100} \Rightarrow$

$60 = \frac{\text{percent} \times 5}{20}$, multiply both sides by 20.

$1200 = \text{percent} \times 5$, divide both sides by 5.

$240 = \text{percent}$; The answer is 240%

4) Answer: C

Let's check the options provided.

A. $-5 + (-18 \div 3) + \frac{-6}{5} \times 5 \rightarrow -5 + (-6) + (-6) = -17$

B. $2 \times (-10) + (-3) \times 4 = (-20) + (-12) = -32$

C. $(-2) + 14 \times 3 \div (-7) = -2 + 42 \div (-7) = -2 - 6 = -8$

D. $(-4) \times (-8) + 5 = 32 + 5 = 37$

WWW.MathNotion.Com

5) Answer: B

Choices A, C and D are incorrect because 60% of each of the numbers is a non-whole number.

A. 61, $\quad 70\% \text{ of } 61 = 0.70 \times 61 = 42.7$

B. 50, $\quad 70\% \text{ of } 50 = 0.70 \times 50 = 35$

C. 42, $\quad 70\% \text{ of } 42 = 0.70 \times 42 = 29.4$

D. 25, $\quad 70\% \text{ of } 25 = 0.70 \times 25 = 17.5$

6) Answer: B

Let x be the integer. Then: $2x - 6 = 80$

Add 8 both sides: $\quad 2x = 86$

Divide both sides by 2: $x = 43$

7) Answer: A

A. $5^3 - 4^3 = 125 - 64 = 61$

B. $2^5 - 2^2 = 32 - 4 = 28$

C. $3^4 - 5^2 = 81 - 25 = 56$

D. $4^4 - 15^2 = 256 - 225 = 31$

8) Answer: C

The radius of the circle is: $\frac{3\pi}{2}$

The area of circle: $\pi r^2 = \pi(\frac{3\pi}{2})^2 = \pi \times \frac{9\pi^2}{4} = \frac{9\pi^3}{4}$

9) Answer: D

Elise has x apple which is 20 apples more than number of apples Alvin owns. Therefore:

$x - 20 = 45 \rightarrow x = 45 + 20 = 65$

Elise has 65 apples.

Let y be the number of apples that Baron has. Then: $y = \frac{1}{5} \times 65 = 13$

10) Answer: 28.

Let x and y be two sides of the shape. Then:

$x + 1 = 1 + 1 + 1 \rightarrow x = 2$

$y + 6 + 2 = 7 + 4 \rightarrow y + 8 = 11 \rightarrow y = 3$

Then, the perimeter is:

$1 + 7 + 1 + 4 + 1 + 2 + 1 + 6 + 2 + 3 = 28$

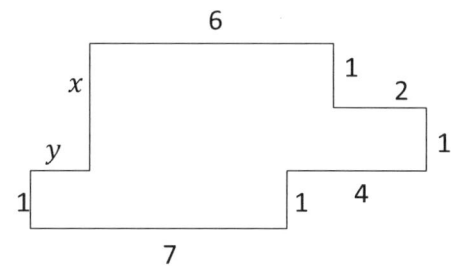

11) Answer: D

Distance that car B travels = 2.2 × distance that car A travels

$= 2.2 \times 341.26 = 750.8$ Km

12) Answer: D

Two months, April and August, in 12 months start with A, then:

Probability $= \dfrac{\text{number of desired outcomes}}{\text{number of total outcomes}} = \dfrac{2}{12} = \dfrac{1}{6}$

13) Answer: D

First, put the numbers in order from least to greatest: 1, 1, 1, 2, 2, 3, 3, 4, 4, 5, 6

The Mode of the set of numbers is: 1 (the most frequent numbers)

Median is: 3 (the number in the middle)

14) Answer: D

If the value of point A is greater than the value of point B, then the distance of two points on the number line is: value of A− value of B

A. $-\dfrac{16}{4} - (-10) = -4 + 10 = 6 = 6$

B. $2 - \left(-\dfrac{16}{4}\right) = 2 + 4 = 6 = 6$

C. $-3 - \left(-\dfrac{16}{4}\right) = -3 + 4 = 1 \neq 6$

15) Answer: 24

$7 + 5 = 12 \implies 144 \div 12 = 12$

$144 \div 12 = 12 \rightarrow 12 \times 5 = 60 \implies 144 - 60 = 84 \implies 84 - 60 = 24$

The ratio of pens to pencils is 5: 7. Therefore there are 5 pens out of all 12 pens and pencils. To find the answer, first dived 144 by 12 then multiply the result by 5.

There are 60 pens and 84 pencils (144-60). Therefore, 24 more pens should be put in the box to make the ratio 1: 1

16) Answer: B

$\frac{35}{8} \cong 4.38$ $\frac{7}{4} = 1.75$ $\frac{9}{12} = 0.75$ $\frac{4}{2} = 2$

Then: $\frac{9}{12} < \frac{7}{4} < \frac{4}{2} < \frac{35}{8}$

17) Answer: B

$3x - 2 = 16 \rightarrow 3x = 16 + 2 = 18 \rightarrow x = \frac{18}{3} = 9$

Then, $3x + 2 = 3(9) + 2 = 27 + 2 = 29$

18) Answer: 3.639

$5(1.153) - 2.126 = 5.765 - 2.126 = 3.639$

19) Answer: 95°

Complementary angles add up to 180 degrees.

$\beta + 160° = 180° \rightarrow \beta = 180° - 160° = 20°$

The sum of all angles in a triangle is 180 degrees. Then:

$\alpha + \beta + 65° = 180° \rightarrow \alpha + 20° + 65° = 180°$

$\rightarrow \alpha + 85° = 180° \rightarrow \alpha = 180° - 85° = 95°$

20) Answer: B

The shaft rotates 360 times in 9 seconds. Then, the number of rotates in 16 second equals to:

$\frac{360 \times 16}{9} = 640$

Practice Test 2

Georgia Milestones Assessment

System - Mathematics

Answers and Explanations

1) Answer: A

For one hour he earns $30, then for t hours he earns $30t. If he wants to earn at least $250, therefor, the number of working hours multiplied by 30 must be equal to 250 or more than 250. $30t \geq 250$

2) Answer: 29

Plug in the value of x and y and use order of operations rule.

$x = 1$ and $y = -2$

$4(3x - 2y) + (4 - 5x)^2 = 4(3(1) - 2(-2)) + (4 - 5(1))^2 = 4(3 + 4) + (-1)^2 = 28 + 1 = 29$

3) Answer: C

$\dfrac{415}{9} \cong 46.11 \cong 46.1$

4) Answer: C

$6(10x - 12) = (6 \times 10x) - (6 \times 12) = (6 \times 10)x - (6 \times 12) = 60x - 72$

5) Answer: C

5% of the volume of the solution is alcohol. Let x be the volume of the solution.

Then: 5% of $x = 40$ ml $\Rightarrow 0.05\ x = 40 \Rightarrow x = 40 \div 0.05 = 800$

6) Answer: A

The coordinate plane has two axes. The vertical line is called the y-axis and the horizontal is called the x-axis. The points on the coordinate plane are address using the form (x, y). The point A is one unit on the left side of x-axis; therefore, its x value is 3 and it is two units up, therefore its y axis is -2. The coordinate of the point is: $(3, -2)$

GMAS Math Practice Tests – Grade 6

7) Answer: A

84: 12 = 14: 2

14 × 6 = 84   And   2 × 6 = 12

8) Answer: −16, 0

Opposite number of any number x is a number that if added to x, the result is 0. Then:

16 + (−16) = 0 and 0 + 0 = 0

9) Answer: A

−50 = 85 − x

First, subtract 85 from both sides of the equation. Then:

−50 − 85 = 85 − 85 − x → −135 = −x

Multiply both sides by (−1) → x = 135

10) Answer: D

Solve for x. −2 ≤ 2x − 4 < 8 ⇒ (add 4 all sides) −2 + 4 ≤ 2x − 4 + 4 < 8 + 4

⇒ 2 ≤ 2x < 12 ⇒ (divide all sides by 2) 1 ≤ x < 6

x is between 1 and 6. Choice D represent this inequality.

11) Answer: D

The ratio of boy to girls is 5:7. Therefore, there are 5 boys out of 12 students. To find the answer, first divide the total number of students by 12, then multiply the result by 5.

900 ÷ 12 = 75 ⇒ 75 × 5 = 375

12) Answer: C

(75 + 5) ÷ (16) = (80) ÷ (16)

The prime factorization of 80 is: 2 × 2 × 2 × 5

The prime factorization of 16 is: 4 × 4

Therefore: (80) ÷ (16) = (2 × 2 × 2 × 5) ÷ (4 × 4)

13) Answer: B

The slope of the line is: $\frac{y_2-y_1}{x_2-x_1} = \frac{9-5}{4-0} = \frac{4}{4} = 1$

The equation of a line can be written as:

$y - y_0 = m(x - x_0)$ → $y - 5 = 1(x - 0)$ → $y - 5 = 1x$ → $y = x + 5$

14) **Answer: 240**

Volume of a box = length × width × height = 8 × 6 × 5 = 240

15) **Answer: B**

Probability = $\frac{number\ of\ desired\ outcomes}{number\ of\ total\ outcomes} = \frac{18}{16+18+14+12} = \frac{18}{60} = \frac{3}{10}$

16) **Answer: A**

Prime factorizing of $30 = 2 \times 3 \times 5$

Prime factorizing of $18 = 3 \times 2 \times 3$

LCM= $2 \times 3 \times 3 \times 5 = 90$

17) **Answer: A**

In any rectangle, both diagonals have equal measure.

18) **Answer: C**

Let's compare each fraction: $\frac{3}{4} < \frac{4}{5} < \frac{26}{29} < \frac{11}{5}$

Only choice C provides the right order.

19) **Answer: D**

$3 \times \frac{5}{7} = \frac{15}{7} = 2.14$

A. $2.14 > 2$

B. $2.14 < 3$

C. $\frac{77}{25} = 3.08 \neq 2.4$

D. $2 < 2.14 < 2.3$ This is the answer!

20) **Answer: 80**

Since, E is the midpoint of AB, then the area of all triangles DAE, DEF, CFE and CBE are equal.

Let x be the area of one of the triangles, then:

$4x = 160 \rightarrow x = 40$

The area of DEC = $2x = 2(40) = 80$

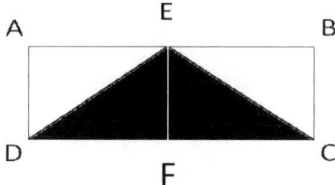

Practice Test 3
Georgia Milestones Assessment
System - Mathematics
Answers and Explanations

1) Answer: B

Plugin the value of x in the equations. $x = -3$, then:

A. $x(2x - 4) = 18 \to -3(2(-3) - 4) = -3(-6 - 4) = -3(-10) = 30 \neq 18$

B. $5(12 - x^2) = 15 \to 5(12 - (-3)^2) = 5(12 - 9) = 5(3) = 15 = 15$

C. $2(-x + 5) = 21 \to 2(-(-3) + 5) = 2(8) = 16 \neq 21$

D. $x(-2x - 11) = -19 \to -3(-2(-3) - 11) = -3(6 - 11) = 15 \neq -19$

2) Answer: D

Use Pythagorean theorem to find the hypotenuse of the triangle.

$a^2 + b^2 = c^2 \to 3^2 + 4^2 = c^2 \to 9 + 16 = c^2 \to 25 = c^2 \to c = 5$

The perimeter of the triangle is: $3 + 4 + 5 = 12$

3) Answer: 250%

$30 = \frac{\text{percent}}{100} \times 12 \Rightarrow 30 = \frac{\text{percent} \times 12}{100} \Rightarrow$

$30 = \frac{\text{percent} \times 3}{25}$, multiply both sides by 25.

$750 = \text{percent} \times 3$, divide both sides by 3.

$250 = \text{percent}$; The answer is 250%

4) Answer: C

Let's check the options provided.

A. $-4 + (-18 \div 3) + \frac{-9}{5} \times 5 \to -4 + (-6) + (-9) = -19$

B. $6 \times (-4) + (-3) \times 4 = (-24) + (-12) = -36$

C. $(-6) + 10 \times 3 \div (-5) = -6 + 30 \div (-5) = -6 - 6 = -12$

D. $(-4) \times (-6) + 11 = 24 + 11 = 36$

5) Answer: B

Choices A, C and D are incorrect because 20% of each of the numbers is a non-whole number.

A. 46, $20\% \text{ of } 46 = 0.20 \times 46 = 9.2$

B. 35, $20\% \text{ of } 35 = 0.20 \times 35 = 7$

C. 52, $20\% \text{ of } 52 = 0.20 \times 52 = 10.4$

D. 28, $20\% \text{ of } 28 = 0.20 \times 28 = 5.6$

6) Answer: B

Let x be the integer. Then: $3x - 5 = 70$

Add 5 both sides: $3x = 75$

Divide both sides by 3: $x = 25$

7) Answer: A

A. $7^3 - 6^3 = 343 - 216 = 127$

B. $2^5 - 5^2 = 32 - 25 = 7$

C. $3^4 - 7^2 = 81 - 49 = 32$

D. $13^2 - 5^3 = 169 - 125 = 44$

8) Answer: A

The radius of the circle is: $\frac{3\pi}{2}$

The area of circle: $\pi r^2 = \pi(\frac{3\pi}{2})^2 = \pi \times \frac{9\pi^2}{4} = \frac{9\pi^3}{4}$

9) Answer: C

Elise has x apple which is 20 apples more than number of apples Alvin owns.

Therefore:

$x - 20 = 40 \rightarrow x = 40 + 20 = 60$

Elise has 60 apples.

Let y be the number of apples that Baron has. Then: $y = \frac{1}{3} \times 60 = 20$

10) Answer: 52.

Let x and y be two sides of the shape. Then:

$x + 3 = 3 + 3 + 3 \to x = 6$

$y + 9 + 5 = 10 + 7 \to y + 14 = 17 \to y = 3$

Then, the perimeter is:

$3 + 10 + 3 + 7 + 3 + 5 + 3 + 9 + 6 + 3 = 52$

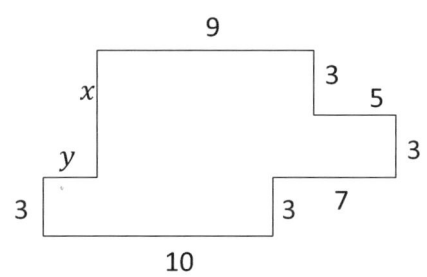

11) Answer: D

Distance that car B travels = 2.5 × distance that car A travels

= 2.5 × 250.50 = 626.25 Km

12) Answer: D

One month, October in 12 months start with O, then:

Probability = $\frac{number\ of\ desired\ outcomes}{number\ of\ total\ outcomes} = \frac{1}{12}$

13) Answer: D

First, put the numbers in order from least to greatest: 3, 4, 4, 5, 5, 5, 6, 6, 7, 8, 9

The Mode of the set of numbers is: 5 (the most frequent numbers)

Median is: 5 (the number in the middle)

14) Answer: D

If the value of point A is greater than the value of point C, then the distance of two points on the number line is: value of A− value of C

A. $-\frac{21}{7} - (-9) = -3 + 9 = 6 = 6$

B. $-3 - \left(-\frac{21}{7}\right) = -3 + 3 = 0 \neq 6$

C. $3 - \left(-\frac{21}{6}\right) = 3 + 3 = 6 = 6$

15) Answer: 39

$2 + 5 = 7 \Rightarrow 91 \div 7 = 13$

$91 \div 7 = 13 \to 13 \times 2 = 26 \Rightarrow 91 - 26 = 65 \Rightarrow 65 - 26 = 39$

The ratio of pens to pencils is 2: 5. Therefore there are 2 pens out of all 7 pens and pencils. To find the answer, first dived 91 by 7 then multiply the result by 2.

There are 26 pens and 65 pencils (91−26). Therefore, 39 more pens should be put in the box to make the ratio 1: 1

16) Answer: C

$\frac{38}{6} \cong 6.33$ $\qquad$ $\frac{8}{3} = 2.66$ $\qquad$ $\frac{7}{9} \cong 0.78$ $\qquad$ $\frac{13}{4} = 3.25$

Then: $\frac{7}{9} < \frac{8}{3} < \frac{13}{4} < \frac{38}{6}$

17) Answer: B

$6x - 2 = 22 \rightarrow 6x = 22 + 2 = 24 \rightarrow x = \frac{24}{6} = 4$

Then, $5x + 5 = 5(4) + 5 = 20 + 5 = 25$

18) Answer: 3.25

$4(1.75) - 3.75 = 7 - 3.75 = 3.25$

19) Answer: 70°

Complementary angles add up to 180 degrees.

$\beta + 125° = 180° \rightarrow \beta = 180° - 125° = 55°$

The sum of all angles in a triangle is 180 degrees. Then:

$\alpha + \beta + 55° = 180° \rightarrow \alpha + 55° + 55° = 180°$

$\rightarrow \alpha + 110° = 180° \rightarrow \alpha = 180° - 110° = 70°$

20) Answer: B

The shaft rotates 330 times in 9 seconds. Then, the number of rotates in 12 second equals to:

$\frac{330 \times 12}{9} = 440$

Practice Test 4

Georgia Milestones Assessment

System - Mathematics

Answers and Explanations

1) Answer: A

For one hour he earns $50, then for t hours he earns $50t. If he wants to earn at least $520, therefor, the number of working hours multiplied by 50 must be equal to 520 or more than 520. $50t \geq 520$

2) Answer: 26

Plug in the value of x and y and use order of operations rule.

$x = 1$ and $y = -2$

$2(3x - y) + (7 - 3x)^2 = 2(3(1) - (-2)) + (7 - 3(1))^2 = 2(3 + 2) + (4)^2 = 10 + 16 = 26$

3) Answer: C

$\dfrac{155}{6} \cong 25.83 \cong 26$

4) Answer: C

$9(10x - 6) = (9 \times 10x) - (9 \times 6) = (9 \times 10)x - (9 \times 6) = 90x - 54$

5) Answer: C

7% of the volume of the solution is alcohol. Let x be the volume of the solution.

Then: 7% of x = 42 ml $\Rightarrow$ 0.07 x = 42 $\Rightarrow$ x = 42 ÷ 0.07 = 600

6) Answer: A

The coordinate plane has two axes. The vertical line is called the y-axis and the horizontal is called the x-axis. The points on the coordinate plane are address using the form (x, y). The point A is three unit on the right side of x-axis; therefore, its x value is 3 and it is three unit down, therefore its y axis is -3. The coordinate of the point is: $(3, -3)$

GMAS Math Practice Tests – Grade 6

7) Answer: A

90: 15 = 30: 5

30 × 3 = 90   And   5 × 3 = 15

8) Answer: $-17, -3$

Opposite number of any number x is a number that if added to x, the result is 0. Then:

$17 + (-17) = 0$ and $3 + (-3) = 0$

9) Answer: A

$-32 = 74 - x$

First, subtract 74 from both sides of the equation. Then:

$-32 - 74 = 74 - 74 - x \rightarrow -106 = -x$

Multiply both sides by $(-1) \rightarrow x = 106$

10) Answer: C

Solve for x. $1 \leq 2x - 5 < 11 \Rightarrow$ (add 5 all sides) $1 + 5 \leq 2x - 5 + 5 < 11 + 5$

$\Rightarrow 6 \leq 2x < 16 \Rightarrow$ (divide all sides by 2) $3 \leq x < 8$

x is between 3 and 8. Choice C represent this inequality.

11) Answer: D

The ratio of boy to girls is 4:5. Therefore, there are 4 boys out of 9 students. To find the answer, first divide the total number of students by 9, then multiply the result by 4.

$450 \div 9 = 50 \Rightarrow 50 \times 4 = 200$

12) Answer: C

$(80 + 4) \div (35) = (84) \div (35)$

The prime factorization of 84 is: $2 \times 2 \times 3 \times 7$

The prime factorization of 35 is: 5×7

Therefore: $(84) \div (35) = (2 \times 2 \times 3 \times 7) \div (5 \times 7)$

13) Answer: B

The slope of the line is: $\frac{y_2 - y_1}{x_2 - x_1} = \frac{7-2}{4-3} = \frac{5}{1} = 5$

The equation of a line can be written as:

$$y - y_0 = m(x - x_0) \rightarrow y - 2 = 5(x - 3) \rightarrow y - 2 = 5x - 15 \rightarrow y = 5x - 13$$

14) Answer: 288

Volume of a box = length × width × height = 8 × 6 × 6 = 288

15) Answer: B

$$\text{Probability} = \frac{number\ of\ desired\ outcomes}{number\ of\ total\ outcomes} = \frac{25}{16+25+20+19} = \frac{25}{80} = \frac{5}{16}$$

16) Answer: A

Prime factorizing of $15 = 3 \times 5$

Prime factorizing of $50 = 2 \times 5 \times 5$

LCM= $2 \times 3 \times 5 \times 5 = 150$

17) Answer: D

In any square, all the statements are true.

18) Answer: C

Let's compare each fraction: $\frac{2}{5} < \frac{6}{7} < \frac{19}{17} < \frac{23}{6}$

Only choice C provides the right order.

19) Answer: D

$3 \times \frac{6}{7} = \frac{18}{7} = 2.57$

A. $2.57 < 4$

B. $2.57 < 3$

C. $\frac{35}{15} = 2.33 \neq 2.57$

D. $2 < 2.57 < 3$ This is the answer!

20) Answer: 120

Since, E is the midpoint of AB, then the area of all triangles DAE, DEF, CFE and CBE are equal.

Let x be the area of one of the triangles, then:

$4x = 240 \rightarrow x = 60$

The area of DEC $= 2x = 2(60) = 120$

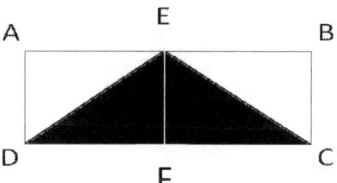

Practice Test 5
Georgia Milestones Assessment
System - Mathematics
Answers and Explanations

1) Answer: B

Plugin the value of x in the equations. $x = -2$, then:

A. $x(3x - 2) = 32 \to -2(3(-2) - 2) = -2(-6 - 2) = -2(-8) = 16 \neq 32$

B. $2(10 - x^2) = 12 \to 2(10 - (-2)^2) = 2(10 - 4) = 2(6) = 12 = 12$

C. $3(-x + 3) = 18 \to 3(-(-2) + 3) = 3(5) = 15 \neq 18$

D. $x(-4x - 10) = -17 \to -2(-4(-2) - 10) = -2(8 - 10) = 4 \neq -17$

2) Answer: D

Use Pythagorean theorem to find the hypotenuse of the triangle.

$a^2 + b^2 = c^2 \to 6^2 + 8^2 = c^2 \to 36 + 81 = c^2 \to 100 = c^2 \to c = 10$

The perimeter of the triangle is: $6 + 8 + 10 = 24$

3) Answer: 300%

Use percent formula: $\text{Part} = \frac{\text{percent}}{100} \times \text{whole}$

$45 = \frac{\text{percent}}{100} \times 15 \Rightarrow 45 = \frac{\text{percent} \times 15}{100} \Rightarrow$

$45 = \frac{\text{percent} \times 3}{20}$, multiply both sides by 20.

$900 = \text{percent} \times 3$, divide both sides by 3.

$300 = \text{percent}$; The answer is 300%

4) Answer: C

Let's check the options provided.

A. $-3 + (-16 \div 4) + \frac{-8}{7} \times 7 \to -3 + (-4) + (-8) = -15$

B. $3 \times (-9) + (-2) \times 5 = (-27) + (-10) = -37$

C. $(-3) + 12 \times 4 \div (-8) = -3 + 48 \div (-8) = -3 - 6 = -9$

D. $(-5) \times (-9) + 36 = 45 + 36 = 81$

5) Answer: B

Choices A, C and D are incorrect because 60% of each of the numbers is a non-whole number.

A. 56, $60\% \ of \ 56 = 0.60 \times 56 = 33.6$

B. 25, $60\% \ of \ 25 = 0.60 \times 25 = 15$

C. 32, $60\% \ of \ 32 = 0.60 \times 32 = 19.2$

D. 48, $60\% \ of \ 48 = 0.60 \times 48 = 28.8$

6) Answer: B

Let x be the integer. Then: $2x - 8 = 90$

Add 8 both sides: $2x = 98$

Divide both sides by 2: $x = 49$

7) Answer: A

A. $6^3 - 5^3 = 216 - 125 = 91$

B. $3^4 - 4^2 = 81 - 16 = 65$

C. $4^3 - 6^2 = 64 - 36 = 65$

D. $6^3 - 12^2 = 216 - 144 = 72$

8) Answer: A

The radius of the circle is: $\frac{8\pi}{2}$

The area of circle: $\pi r^2 = \pi(\frac{8\pi}{2})^2 = \pi \times \frac{64\pi^2}{4} = 16\pi^3$

9) Answer: C

Elise has x apple which is 30 apples more than number of apples Alvin owns.

Therefore:

$x - 30 = 50 \rightarrow x = 50 + 30 = 80$

Elise has 80 apples.

Let y be the number of apples that Baron has. Then: $y = \frac{1}{4} \times 80 = 20$

10) Answer: 56.

Let x and y be two sides of the shape. Then:

$x + 2 = 2 + 2 + 2 \rightarrow x = 4$

$y + 12 + 4 = 14 + 8 \rightarrow y + 16 = 22 \rightarrow y = 6$

Then, the perimeter is:

$2 + 14 + 2 + 8 + 2 + 4 + 2 + 12 + 4 + 6 = 56$

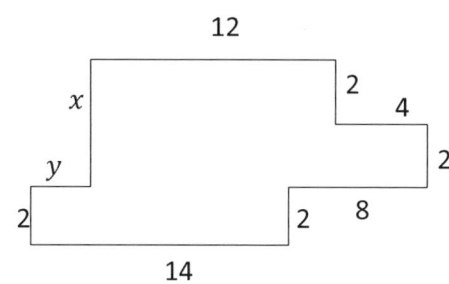

11) Answer: D

Distance that car B travels = 3.2 × distance that car A travels

= 3.2 × 241.15 = 771.68 Km

12) Answer: D

Two months, March and May, in 12 months start with M, then:

Probability = $\frac{number\ of\ desired\ outcomes}{number\ of\ total\ outcomes} = \frac{2}{12} = \frac{1}{6}$

13) Answer: D

First, put the numbers in order from least to greatest: 1, 2, 3, 3, 4, 5, 5, 6, 6, 6, 8

The Mode of the set of numbers is: 6 (the most frequent numbers)

Median is: 5 (the number in the middle)

14) Answer: D

If the value of point A is greater than the value of point B, then the distance of two points on the number line is: value of A− value of B

A. $-\frac{12}{3} - (-12) = -4 + 12 = 8 = 8$

B. $4 - \left(-\frac{12}{3}\right) = 4 + 4 = 8 = 8$

C. $-6 - \left(-\frac{12}{3}\right) = -6 + 4 = -2 \neq 8$

15) Answer: 14

$3 + 4 = 7 \Rightarrow 98 \div 7 = 14$

$98 \div 7 = 14 \rightarrow 14 \times 3 = 42 \Rightarrow 98 - 42 = 56 \Rightarrow 56 - 42 = 14$

The ratio of pens to pencils is 3: 4. Therefore there are 3 pens out of all 7 pens and pencils. To find the answer, first dived 98 by 7 then multiply the result by 3.

There are 42 pens and 56 pencils (98-42). Therefore, 14 more pens should be put in the box to make the ratio 1: 1

16) Answer: C

$\frac{45}{8} = 5.63$ $\quad\quad$ $\frac{5}{2} = 2.5$ $\quad\quad$ $\frac{8}{15} \cong 0.533$ $\quad\quad$ $\frac{9}{3} = 3$

Then: $\frac{8}{15} < \frac{5}{2} < \frac{9}{3} < \frac{45}{8}$

17) Answer: B

$4x - 3 = 17 \rightarrow 4x = 17 + 3 = 20 \rightarrow x = \frac{20}{4} = 5$

Then, $4x + 3 = 4(5) + 3 = 20 + 3 = 23$

18) Answer: 5.393

$3(2.313) - 1.546 = 6.939 - 1.546 = 5.393$

19) Answer: 70°

Complementary angles add up to 180 degrees.

$\beta + 145° = 180° \rightarrow \beta = 180° - 145° = 35°$

The sum of all angles in a triangle is 180 degrees. Then:

$\alpha + \beta + 75° = 180° \rightarrow \alpha + 35° + 75° = 180°$

$\rightarrow \alpha + 110° = 180° \rightarrow \alpha = 180° - 110° = 70°$

20) Answer: B

The shaft rotates 440 times in 11 seconds. Then, the number of rotates in 15 second equals to:

$\frac{440 \times 15}{11} = 600$

Practice Test 6

Georgia Milestones Assessment

System - Mathematics

Answers and Explanations

1) Answer: A

For one hour he earns $40, then for t hours he earns $40t. If he wants to earn at least $450, therefor, the number of working hours multiplied by 40 must be equal to 450 or more than 450. $40t \geq 450$

2) Answer: 37

Plug in the value of x and y and use order of operations rule.

$x = 2$ and $y = -1$

$2(2x - 2y) + (3 - 4x)^2 = 2(2(2) - 2(-1)) + (3 - 4(2))^2 = 2(4 + 2) + (-5)^2 = 12 + 25 = 37$

3) Answer: C

$\dfrac{369}{7} \cong 52.71 \cong 52.7$

4) Answer: C

$8(11x - 9) = (8 \times 11x) - (8 \times 9) = (8 \times 11)x - (8 \times 9) = 88x - 72$

5) Answer: C

6% of the volume of the solution is alcohol. Let x be the volume of the solution.

Then: 6% of $x = 36$ ml $\Rightarrow 0.06\ x = 36 \Rightarrow x = 36 \div 0.06 = 600$

6) Answer: A

The coordinate plane has two axes. The vertical line is called the y-axis and the horizontal is called the x-axis. The points on the coordinate plane are address using the form (x, y). The point A is one unit on the left side of x-axis; therefore, its x value is 1 and it is three units down, therefore its y axis is -3. The coordinate of the point is: $(1, -3)$

WWW.MathNotion.Com

7) Answer: A

36: 12 = 12: 4

$12 \times 3 = 36$ And $4 \times 3 = 12$

8) Answer: $-22, 0$

Opposite number of any number x is a number that if added to x, the result is 0. Then:

$22 + (-22) = 0$ and $0 + 0 = 0$

9) Answer: A

$-42 = 94 - x$

First, subtract 94 from both sides of the equation. Then:

$-42 - 94 = 94 - 94 - x \rightarrow -136 = -x$

Multiply both sides by $(-1) \rightarrow x = 136$

10) Answer: D

Solve for x. $-3 \leq 3x - 6 < 9 \Rightarrow$ (add 6 all sides) $-3 + 6 \leq 3x - 6 + 6 < 9 + 6$

$\Rightarrow 3 \leq 3x < 15 \Rightarrow$ (divide all sides by 3) $1 \leq x < 5$

x is between 1 and 5. Choice D represent this inequality.

11) Answer: D

The ratio of boy to girls is 3:4. Therefore, there are 3 boys out of 7 students. To find the answer, first divide the total number of students by 7, then multiply the result by 3.

$700 \div 7 = 100 \Rightarrow 100 \times 3 = 300$

12) Answer: C

$(45 + 5) \div (25) = (50) \div (25)$

The prime factorization of 50 is: $2 \times 5 \times 5$

The prime factorization of 25 is: 5×5

Therefore: $(50) \div (25) = (2 \times 5 \times 5) \div (5 \times 5)$

13) Answer: B

The slope of the line is: $\frac{y_2 - y_1}{x_2 - x_1} = \frac{4-2}{3-1} = \frac{2}{2} = 1$

The equation of a line can be written as:

$y - y_0 = m(x - x_0) \rightarrow y - 2 = 1(x - 1) \rightarrow y - 2 = x - 1 \rightarrow y = x + 1$

14) Answer: 150

Volume of a box = length × width × height = 10 × 5 × 3 = 150

15) Answer: B

Probability = $\frac{number\ of\ desired\ outcomes}{number\ of\ total\ outcomes} = \frac{20}{13+20+15+17} = \frac{20}{65} = \frac{4}{13}$

16) Answer: A

Prime factorizing of 40 = 2 × 2 × 2 × 2 × 5

Prime factorizing of 24 = 2 × 2 × 2 × 2 × 3

LCM = 2 × 2 × 2 × 2 × 3 × 5 = 120

17) Answer: D

In any square, all the statements are true.

18) Answer: C

Let's compare each fraction: $\frac{3}{5} < \frac{5}{7} < \frac{27}{30} < \frac{13}{6}$

Only choice C provides the right order.

19) Answer: D

$4 \times \frac{7}{11} = \frac{28}{11} = 2.54$

A. 2.54 > 2

B. 2.54 < 4

C. $\frac{77}{25} = 3.08 \neq 3.11$

D. 2 < 2.54 < 3   This is the answer!

20) Answer: 60

Since, E is the midpoint of AB, then the area of all triangles DAE, DEF, CFE and CBE are equal.

Let x be the area of one of the triangles, then:

$4x = 120 \rightarrow x = 30$

The area of DEC = $2x = 2(30) = 60$

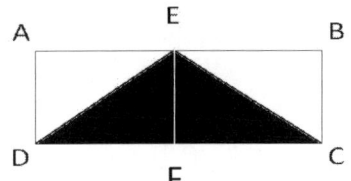

"End"

Made in the USA
Columbia, SC
15 April 2025

56638914R00057